by Kathleen Powell

SCHOOL PUBLISHERS

Orlando   Austin   New York   San Diego   Toronto   London

Visit *The Learning Site!*
www.harcourtschool.com

A Day Without Electricity

One day last week, the Jenks sisters, Janet and Sylvia, were on their way home from school. Sylvia shivered. "I'm cold! When I get home, I'm going to have a cup of hot chocolate and sit next to the heater."

Janet said, "Hot chocolate? I'd love some, too. And I'm going to watch my favorite TV show."

"You can't do that!" said Sylvia. "I have to listen to a CD for my music class."

When they got home, they yelled, "Hi, Mom, we're home! Can we have some hot chocolate?"

Janet said, "Can I watch TV?"

And Sylvia said, "I have to listen to my CD."

Mrs. Jenks hugged them both. She said, "There won't be any hot chocolate, at least not for a while. And we can't watch TV or listen to a CD, either. The electric company is working on the power lines. We don't have electricity."

"Oh no," the girls groaned.

"I guess I'll get a cookie and sit beside the heater," said Sylvia. "I'm cold."

Janet said, "I'll play a game in the basement."

But their mom told them that without electricity, the heater wouldn't work and the basement lights wouldn't turn on.

"Without electricity, we can't do anything," said Janet.

"The power company said that the electricity should come back on in about an hour. Why don't we sit on the couch and tell stories? Sylvia, run and get us some blankets."

When they had settled in on the couch, Janet started listing all of the things that don't work without electricity: the lights, the heater, the CD player, the TV, and the stove. Mrs. Jenks added the radio, the microwave, the dishwasher, the iron, the toaster, and the hairdryer. Sylvia said, "Wow. Where does all of that electricity come from?"

Where Electricity Comes From

Mrs. Jenks explained that the electricity in their home comes from a generating plant. She said that there are different kinds of electric stations, but the one that supplies their town generates electricity by burning coal.

She told them that electricity generated by the plant moves through wires from the plant to their home. When they plug in the heater or the CD player, the plug makes contact with wires from the plant. Electricity from the generating plant then moves into the heater or the CD player.

Coal-burning generating station

She said that electrical energy can come from batteries, too. Batteries contain chemicals. When you put batteries in a flashlight, the chemicals cause electricity to flow in the flashlight.

Janet asked whether she should get the flashlight, just in case it got dark before the power came back on. They decided that that would be a good idea. Janet went to the kitchen and got the flashlight. When she got back, she asked, "How does electricity make light?"

Light from Electricity

Mrs. Jenks explained that electricity produces light in a light bulb. There are two kinds of light bulbs: incandescent bulbs and fluorescent bulbs.

Incandescent light bulb

"In an incandescent light bulb, electricity flows through a wire in the bulb. The wire is called a filament. The filament is so thin that electricity can't flow through it easily. This causes the wire to get hot. The wire gets so hot that it glows. The glowing filament is what you see as light."

"Fluorescent bulbs don't have filaments. They are filled with a gas that glows when electricity flows into them. This light then causes a chemical on the inside of the bulb to glow." She asked Janet to look at the flashlight and see whether it had an incandescent bulb or a fluorescent bulb. Sylvia helped Janet open up the flashlight and take the bulb out. Janet said, "It's an incandescent bulb. I can see the wire filament."

Heat from Electricity

"Okay," said Sylvia, "I see how electricity produces light, but how does it produce heat?"

Fluorescent light bulb

◀ Electricity flows through wires in a heater.

Mrs. Jenks said, "Remember how the wire in an incandescent light bulb heats up? In a heater, electricity also flows through wires and makes them heat up. They light up, too. But the heater uses a different kind of wire, so when electricity goes through it, it produces a lot of heat and only a little bit of light."

Sound from Electricity

Sylvia said, "What about my CD player? How does electricity make the sound I hear when I play my CDs?"

Mrs. Jenks said, "That's a bit more complicated. When a CD is made, a microphone changes sound waves into pulses of electricity. The CD stores these pulses. The CD player reads the pulses and turns them back into electricity. This electricity moves through wires to speakers. Each speaker has a cone-shaped part inside. Electricity and magnets combine to move the cone back and forth quickly. The movements of the cone make sound waves that travel to your ears."

Just then the lights came on. Janet jumped up, "Wow! We've got electricity again. I'm going to watch my TV show."

Mom said, "It's too late. Your show's over."

Janet looked at the clock above the TV. "No, it isn't. The clock says 3:30."

Mom said, "Didn't you notice? The clock's electric, too. When the power went off, it stopped. It's really 4:30. But now that the electricity is back on, I can make all of us some hot chocolate."

Speakers produce sound from electricity. ▶